spot

WILD CATS

BOBCATS

by Alissa Thielges

AMICUS | AMICUS INK

tail

markings

Look for these words and pictures as you read.

eyes

paws

A bobcat stands
on a tree.
It is ready to hunt.

Bobcats are twice
as big as house cats.
They live in forests.

tail

See the short tail?

It is bobbed.

The bobcat is named after its tail.

eyes

See the eyes?
They see in the dark.
A bobcat hunts at night.

See the markings?
There are spots and stripes.
They help the bobcat hide.

markings

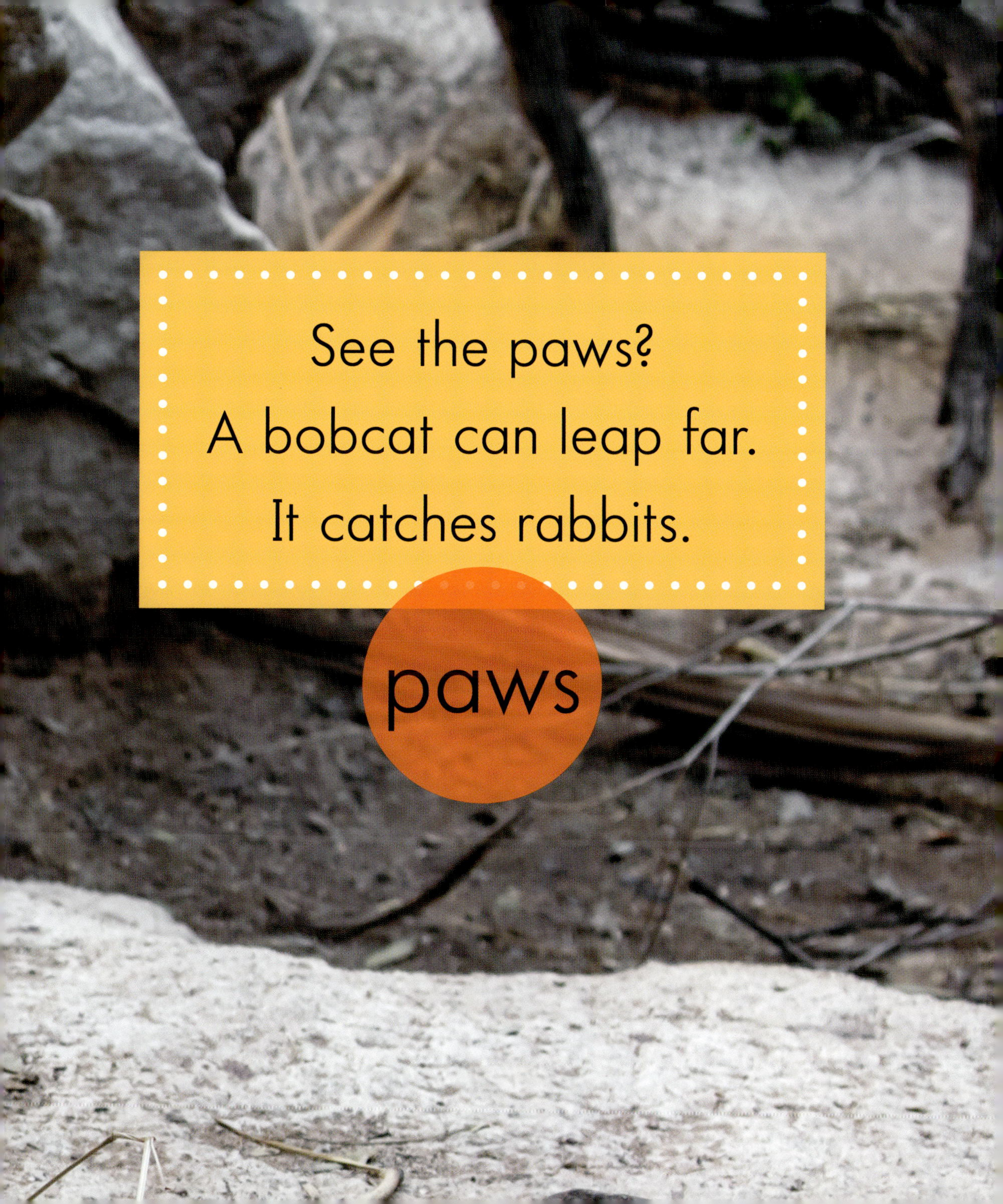

See the paws?
A bobcat can leap far.
It catches rabbits.

paws

A baby bobcat watches mom.
It learns to hunt.

tail

markings

Did you find?

eyes

paws

Spot is published by Amicus and Amicus Ink
P.O. Box 1329, Mankato, MN 56002
www.amicuspublishing.us

Library of Congress Cataloging-in-Publication Data
Names: Thielges, Alissa, 1995- author.
Title: Bobcats / by Alissa Thielges.
Description: Mankato, MN : Amicus/Amicus Ink, [2021] | Series: Spot | Audience: Ages 4-7 | Audience: Grades K-1
Identifiers: LCCN 2019036326 (print) | LCCN 2019036327 (ebook) | ISBN 9781681519272 (library binding) | ISBN 9781681525747 (paperback) | ISBN 9781645490128 (pdf)
Subjects: LCSH: Bobcat--Juvenile literature.
Classification: LCC QL737.C23 T473533 2021 (print) | LCC QL737.C23 (ebook) | DDC 599.75/36--dc23
LC record available at https://lccn.loc.gov/2019036326
LC ebook record available at https://lccn.loc.gov/2019036327

Printed in the United States of America

HC 10 9 8 7 6 5 4 3 2 1
PB 10 9 8 7 6 5 4 3 2 1

Gillia Olson, editor
Deb Miner, series designer
Ciara Beitlich, book designer & photo researcher

Photos by Alamy/Stephen J. Krasemann 6–7; Getty/Louise Heusinkveld 3; Getty/Carol Gray 12–13; iStock/Chuck Schug Photography 8–9; iStock/oneword 14–15; Shutterstock/Evelyn Harrison 4–5; Shutterstock/Warren Metcalf 10–11; Superstock/M. Watsonantheo/Pantheon cover, 16; Shutterstock/Geoffrey Kuchera 1